Russell Discombe

THE BACK GARDEN ASTRONOMER

Photographing the universe from home

Quadrant Books

2nd Floor, 6-8 Dyer Street, Cirencester, Gloucestershire, GL7 2PF
An imprint of Memoirs Books. www.mereobooks.com
and www.memoirsbooks.co.uk

The back garden astronomer

First published in Great Britain by Quadrant Books in 2023

Copyright © Russell Discombe 2023

Russell Discombe has asserted his right under the Copyright Designs and
Patents Act 1988 to be identified as the author of this work.

All rights reserved.
No part of this publication may be reproduced, stored in a retrieval system,
or transmitted in any form or by any means, without the prior permission in
writing of the publisher, nor be otherwise circulated in any form of binding or
cover other than that in which it is published and without a similar condition
including this condition being imposed on the subsequent purchaser

A catalogue record for this book is available from the British Library

Paperback ISBN 978-1-7398645-5-2

Design by Ray Lipscombe
Printed and bound in Great Britain
This is a work of nonfiction.

One of my landscape astro images of the Milky Way taken in Pembrokeshire, a dark sky discovery site offering some of the best dark sky opportunities in the country.

Like many people, I decided to use my time during the Covid lockdown to start a new hobby. I have always loved photography and, alongside my astrophotography, I am a keen amateur wildlife and landscape photographer. Before this year, my only experience of astrophotography was wide angle landscapes which included the stars and the Milky Way.

SOMEWHERE, SOMETHING INCREDIBLE IS WAITING TO BE KNOWN.
Carl Sagan

I was introduced to landscape astrophotography by a friend and fellow photographer, Ben Jarvis, approximately five years ago. Ben invited me on a shoot where I took my first image of the Milky Way. When I saw its faint outline pop up on the back of my camera for the first time, I was instantly hooked. I have been taking landscape astro photos ever since and have travelled to some of the darkest skies across the South West and Wales in pursuit of better images.

In pursuit of a new hobby and something that I could delve into during lockdown, I took to the internet. After watching YouTube videos from some amazing deep sky astrophotographers and doing some research online, I instantly knew that deep sky astrophotography was a hobby I would love. The images I saw taken by amateur photographers blew my mind. I never dreamt that such amazing photographs could be taken from someone's garden. So during the lockdowns of 2020-2021, I decided to turn my attention to the skies.

The hobby combines my understanding of the technical side of photography with my fascination with space. Having read a few articles online and some astrophotography books, I quickly discovered that capturing a deep sky image wasn't as simple as buying a telescope and attaching a camera. There were so many things to consider such as the Earth's rotation, the necessary extremely long exposure times, light pollution, image editing and stacking, finding targets in the sky, auto guiding and filters, to name just a few. Indeed, when I first started the hobby, the books and internet posts I was reading at times felt like a foreign language, with phrases and words I had never heard before.

It soon became obvious that my journey into deep sky astrophotography was going to be a steep learning curve. I decided that the only way I was going to learn was to jump in and attempt my first image. Just after the first official lockdown, I purchased my first star tracker, the Skywatcher Star Adventurer, and started on my journey into the captivating world of deep space astrophotography.

My first set up: the Sony a7iii with the Sony 100-400mm lens on the Skywatcher Star Adventurer.

TRACKING THE STARS

The extra time stuck at home luckily coincided with numerous clear nights during the spring and summer, which gave me the perfect opportunity to learn how my new star tracker worked. I spent every clear night in the garden in pursuit of that first image. Having never used a star tracker before, I unsurprisingly failed for the first few nights.

The concept of equatorial mounts, commonly known as star trackers, is relatively simple. The earth rotates, but the stars do not move. Therefore, if you want to take a long exposure of the stars without star trailing, you need to use a star tracker to counter the Earth's rotation. These star trackers move at the same speed as the Earth but in the opposite direction. For them to work, you need to align them to the Earth's axis by pointing them at Polaris (the North Star). After a few mistakes and again watching more YouTube videos (it's amazing what you can learn on YouTube!), I managed to take my first ever deep space image of Andromeda (see next page).

I took my initial images with a small star tracker, my Sony mirrorless camera and camera lenses. However, over the year I invested in three different telescopes, a larger equatorial mount and multiple dedicated astronomy cameras (both mono and two colour versions) in order to capture a wider range of targets in the night sky. This photo-book shows my progression through my first year and a half of astrophotography and all of the images that you see in this book were taken between March 2020 and September, with the vast majority taken during Covid-19 lockdown restrictions.

The picture (opposite) of the Andromeda Galaxy is my first-ever deep sky image. The book contains a selection of galaxy photos taken with a Sony mirrorless camera followed by a number of galaxy images taken with my astronomy cameras. It then progresses to photos of different nebulae within our Milky Way. Most of these have been taken with narrowband filters in order to isolate different gases and accentuate the nebula, and these are probably my favourite type of image.

Towards the end of the summer of 2020, the lockdown restrictions started to ease. This happened just in time for me to capture the end of the Milky Way core season and allowed me to photograph a few wide-angle landscape images which can be seen towards the end of the book. 2020 also brought a special visitor in comet Neowise, which is now the most photographed comet of all time. You can see my attempt at capturing this incredible comet later in the book. I hope you enjoy the view of our universe from my garden in Cirencester, Gloucestershire, UK and the journey which has taken me from a complete beginner to having a picture published by NASA as the Astronomy Photo of the Day (APOD).

Skywatcher 190MN telescope, ZWO ASI 1600mm camera and filter wheel on the Skywatcher NEQ6-Pro mount.

William Optics Zenithstar 73 (Z73) telescope.

The Andromeda Galaxy (M31)

This is my first ever deep-sky astro image and, while it is technically not perfect, I still love this photo. The Andromeda Galaxy is the nearest large galaxy to ours (the Milky Way) and it is estimated to have 1 trillion stars. It is approximately 2.5 million light-years away, meaning how we see Andromeda now is how it looked 2.5 million years ago. The light captured in this photo took 2.5 million years to reach my camera sensor. This galaxy is believed to be the most distant object that we can see with our naked eye. Due to its large size and brightness, it is an easy target to locate and one of the first targets for most visual astronomers and astrophotographers.

This image was taken with the Sony A7iii and the 100-400mm lens. This images only contains 15 two-minute exposures, making it one of my shortest images to date.

The Pinwheel Galaxy (M101)

The Pinwheel Galaxy is approximately 25 million light-years from Earth in the constellation Ursa Major. The Pinwheel is approximately 170,000 light-years across, making it nearly twice the size of our galaxy, the Milky Way, and it is believed to contain over a trillion stars – more than twice as many as the Milky Way. It is an excellent example of a spiral galaxy and the arms of the galaxy contain numerous star-forming nebulae, causing the bright colours that you see in this image.

This image was taken with the Sony a7iii and the 100-400mm lens. It contains 84 four-minute exposures stacked together in post processing.

On the next two pages, you will see more recent images I have taken of the Andromeda and Pinwheel galaxies. These were taken in February 2021, almost a year after starting the hobby with my upgraded and more specialist equipment.

The Andromeda Galaxy (M31)

There are some targets in the night sky that just keep drawing you back again and again. For me, the Andromeda Galaxy is one of those. I have photographed this target four times now and I imagine that I will head back again. As this galaxy is so close to us (2.5 million light-years away), you can really get some detail in the dust lanes and arms of in this spiral galaxy.

This image was taken with the ZWO ASI 1600mm camera using the LRGB filters and the William Optics Z73 telescope. It contains just over 4 hours of total integration time split evenly between the LRGB filters.

The Pinwheel Galaxy (M101)

Here is my second attempt at photographing the Pinwheel Galaxy, this time with my dedicated astro camera and telescope. The upgraded equipment really brings out the detail and colour in this target. One of the key benefits of the astro camera is that it can pick up the red signal (hydrogen alpha) from the different targets, something that my normal camera cannot do.

This image was taken with my asi2600mc pro and the Skywatcher 190MN telescope. This image contains just over 2.5 hours of total integration time.

The Whirlpool Galaxy (M51)

The Whirlpool Galaxy is approximately 31 million light-years from Earth in the constellation Canes Venatici. The galaxy is approximately half the size of the Milky Way. The 'arms' of this galaxy contain numerous star-forming factories with large amount of hydrogen gas. This image actually contains two galaxies: the Whirlpool Galaxy (M51), a grand-design spiral galaxy, and the small, yellowish section which is actually a galaxy in its own right known as NGC 5195. It was once believed that the two galaxies were colliding, however, the Hubble telescope has revealed that NGC 5195 is actually passing behind the Whirlpool Galaxy.

This image was taken with the Sony a7iii and the William Optics Z73. It was created by stacking 54 four-minute exposures in post processing. On the next page, you will see the same target taken with my astro camera.

The Needle Galaxy (NGC 4656)

This is probably the most distant target I have photographed to date. It is believed to be between 30-50 million light-years from Earth. The light that hit my camera sensor left the Needle Galaxy between 30-50 million years ago, a fact that still blows my mind. Also known as NGC 4565, it is the finest and brightest example of an edge-on spiral galaxy visible to us and my favourite of the galaxy images I have taken so far. It is a huge target, being approximately a third larger than our galaxy the Milky Way Galaxy.

Taken with the Sony A7iii and the Skywatcher 190MN telescope, this image contains 41 four-minute exposures stacked in post processing.

The Whirlpool Galaxy (M51)

The same target as the previous page, this image was taken approximately 11 months after the first using my dedicated astro camera.
This image was taken with the ZWO ASI 1600mm camera using the LRGB filters and the Skywatcher 190MN.

This image contains just 5 hours 30 minutes of total integration time split evenly between the LRGB filters.

The Leo Triplet (M65, M66, NGC 3628)

The Leo Triplet is a small group of galaxies approximately 35 million light-years from Earth in the constellation Leo. The galaxy group consists of the spiral galaxies M65 (top right), M66 (bottom right) and NGC 3628. NGC 3628 is otherwise known as the Hamburger Galaxy, and some think it resembles a hamburger, although I cannot see the likeness. What do you think?

This image was taken with the Sony A7iii and the Skywatcher 190MN telescope. It contains 30 four-minute exposures stacked in post processing.

The Triangulum Galaxy (M33)

The Triangulum Galaxy is a spiral galaxy approximately 2.73 million light-years from Earth. It is the smallest galaxy in our local group, which includes the Milky Way and the Andromeda Galaxy. The biggest of the 'local group' is the Andromeda Galaxy and it is believed that this will one day collide with and consume the Milky Way and the Triangulum Galaxy. Thankfully though, that's not predicted for another 4 billion years or so!

Taken with the ZWO ASI 1600mm camera using the LRGB filters and the William Optics Z73 telescope, this image contains just under 3 hours of total integration time split evenly between the LRGB filters.

"TO CONFINE OUR ATTENTION TO TERRESTRIAL MATTERS WOULD BE TO LIMIT THE HUMAN SPIRIT."

Stephen Hawking

The Fireworks Galaxy (NGC 6946)

This galaxy is an intermediate spiral galaxy with a small bright nucleus, approximately 25.2 million light-years from Earth. The light that hit my camera sensor left the galaxy 25.2 million years ago, so this is what this object looked like millions of years in the past. It is a fairly small galaxy, with a diameter of 40,000 light-years, making it approximately one third the size of the Milky Way.

Taken with the ZWO ASI 1600mm camera using the LRGB filters and the Skywatcher 190MN telescope, this image contains just over 4 hours of total integration time split evenly between the LRGB filters.

Bode's Galaxy and The Cigar Galaxy (M81 and M82)

M81 (Bode's Galaxy) is a spiral galaxy, while M82 (The Cigar Galaxy) is an irregular galaxy. Both are approximately 11.8 million light-years from Earth. They are relatively easy to find in the constellation Ursa Major and can easily be seen through binoculars or a small telescope.

The image was taken with the Skywatcher 190MN and the Sony A7iii. It consists of 59 3½-minute exposures stacked together in post processing.

The Sunflower Galaxy (M 63)

The Sunflower Galaxy is known as a flocculent spiral galaxy. Unlike the well-known grand design spiral galaxies, flocculent spiral galaxies do not have the well-defined spiral arms. Instead, the arms appear to be slightly disjointed. The galaxy is located approximately 27 million light-years from Earth in the constellation Canes Venatici. Although the Sunflower Galaxy is approximately the same size as the Milky Way, the distance from Earth means that a larger telescope is needed to view or photograph this target. With binoculars or a small telescope, the galaxy will only appears as a faint blur or patch of light.

The image was taken with the Skywatcher 190MN and the ZWO ASI2600mc pro astronomy camera. It consists of 39 five-minute exposures stacked together in post processing.

"BE CLEARLY AWARE OF THE STARS AND INFINITY ON HIGH. THEN LIFE SEEMS ALMOST ENCHANTED AFTER ALL."

Vincent van Gogh

The Whale and Hockey Stick Galaxies (NCG 4631 & NGC 4656)

The Whale Galaxy (NGC 4631) is an edge-on barred spiral galaxy in the constellation Canes Venatici. The Galaxy is approximately the same size as our Galaxy, the Milky Way, and about 25 million light-years from Earth. The Hockey Stick Galaxy (NGC 4656) is a nearby galaxy located in the same constellation. This unusual shape of the hockey stick is thought to be due to an interaction between NGC 4656 and a couple of near neighbouring galaxies, including the Whale Galaxy.

The image was taken with the Skywatcher 190MN and the ZWO ASI2600mc pro astronomy camera. It consists of 44 five-minute exposures.

M106 & NGC 4217

M106 is a spiral galaxy located in the constellation Canes Venatici. M106 lies at a distance of 23.7 million light years from Earth and has an approximate diameter of 135,000 light-years, making it larger than the Milky Way. In this frame you can also see NGC 4217 which, is an edge-on spiral galaxy which lies approximately 60 million light years from Earth. I really like the different shapes of these two galaxies together in one image. If you look closely at this image (and many of the other images in this book) you will see a number of smaller, more distant galaxies in the frame. After running the photo through editing software which annotates the image, I discovered that there are 33 visible galaxies in this frame, some of which could be billions of light-years from Earth

This image was taken with the 190MN and the ZWO ASI2600mc pro camera. It consists of 89 five-minute exposures.

Bode Black Eye Galaxy (M64)

M64 is a spiral galaxy located in the constellation of Coma Berenice, approximately 17 million light-years from Earth. This galaxy is often referred to as the "Black Eye" or "Evil Eye" galaxy because of the dark band of dust causing its eye-like appearance. The gas in the extremities of this galaxy is believed to be rotating in the opposite direction from the gas and stars in the centre. This unusual behaviour has been linked to a collision between M64 and a satellite galaxy over one billion years ago.

This image was taken with the ZWO ASI2600mc pro camera and the Skywatcher 190MN telescope. It contains 100 five-minute exposures and is one of my favourite galaxy images of those I have taken.

Hickson 44 Galaxy Group

Hickson 44 is a group of four galaxies in the constellation Leo: NGC 3185, NGC 3187, NGC 3190, and NGC 3193. The group is 60-100 million light-years from Earth and is one of the most distant targets I have photographed.

The two galaxies in the middle of the image are the are edge-on galaxy NGC 3190, right next to the very unusual and distinctive S-shaped galaxy NGC 3187. NGC 3193, towards the bottom right of the group, is a bright elliptical galaxy. The smaller spiral galaxy in the top left corner is NGC 3185, the 4th member of the Hickson group.

This image contains 44 five-minute exposures captured with the asi2600mc Pro and the Skywatcher 190MN telescope. Because this target is so distant, the individual galaxies appear very small in the frame. I didn't collect enough data on this target to really do it justice and bring out the colours and details in the galaxies. I am not very happy with the final image, but when I was writing this book Leo was too low on the horizon to collect any more data to add to the image. I will therefore put this target on my 'to photograph' list for next spring.

M109

M109 is a barred spiral galaxy in the constellation Ursa Major. It is approximately 83.5 million light-years from Earth meaning along with the Hickson 44 galaxy group is one of the most distant targets I have photographed. The light that hit my camera sensor left the target 83.5 million years ago. These types of distance always astound me.

This image was taken with the ZWO ASI2600mc pro and the Skywatcher 190MN telescope. The image contains 55 five-minute exposures.

The Sombrero Galaxy (M104)

The Sombrero Galaxy is a striking edge-on spiral galaxy located 28 million light-years away in the constellation Virgo. The famous Hubble Telescope image of this galaxy made it one of the most recognisable and most popular targets for amateur astronomers. It is believed that the galaxy is moving away from us at a rate of 700 miles per second and the discovery of this incredible speed provided some of the earliest evidence that the universe was expanding in all directions. The location of this galaxy (in Virgo) is an exceptionally hard part of the night sky for me to photograph as my house blocks a large part of the view to the south. I only had a very short window (less than a month) to capture data on this target and within this window I only had two clear nights before it was no longer visible. I did not therefore manage to capture as much data as I would have liked to really bring out the detail. It's another target that I will add to my list for next year.

This image was taken with the ZWO ASI2600mc pro and the Skywatcher 190NM telescope. It contains 38 five-minute exposures.

Owl Nebula and Surfboard Galaxy (M97 & M108)

A galaxy and a nebula in one image. Both these targets can be found in the constellation Ursa Major and with careful framing can fit into a single image at 1000mm. I probably should have used my wider 430mm telescope for a better framing of these targets. The Owl Nebula is a planetary nebula approximately 2030 light-years from Earth and it is believed that the nebula is 8000 years old. The surfboard galaxy is approximately 45.9 million light-years from Earth.

This image was taken with the ZWO ASI2600mc pro and the Skywatcher 190NM telescope. It contains 55 five-minute exposures.

The Great Globular Cluster in Hercules (M13)

When there is a full moon the quality of data you can collect really suffers. The moon acts as a huge source of light pollution, making it incredibly difficult to capture images. Fortunately one thing you can photograph is globular clusters. A globular cluster is a spherical cluster of stars on the outskirts of a galaxy which is very tightly bound by gravity. M13 (pictured here) is described as the finest example of a globular cluster in the northern hemisphere. In this cluster there are over 100,000 stars, some of which are believed to be 12-13 billion years old – nearly as old as the universe itself.

This image was taken with the ZWO ASI2600mc pro and the Skywatcher 190NM telescope. It contains 54 one-minute exposures.

The Elephant's Trunk Nebula (IC 1396)

The remaining deep sky images are all nebulae and most have been taken with my dedicated mono astronomy camera (the ZWO ASI 1600mm) and narrowband filters. The camera is a mono (black and white) camera and is extremely sensitive to light. As it is a mono camera, you need to shoot through different filters and assign the filters to a colour channel to create a colour image. The most popular way of doing this is to shoot through narrowband filters which isolate Hydrogen Alpha (Ha), Oxygen 3 (Oiii) and Sulphur 2 (Sii) gases and assigning them to different colour channels. I like the effect that assigning Ha to green, Sii to red and Oiii to blue gives you. This is known as the SHO or the Hubble palette as it is the approach that the Hubble team used to produce a lot of their images.

Located in the constellation Cepheus, IC 1396 and the Elephant's Trunk Nebula is located approximately 2400 light-years from Earth. This beautiful area of hydrogen gas and dust is a stellar nursery (star-forming site) that holds many young stars. The top region of the Elephant's Trunk is being blown away by radiation emitted from new-born stars that are igniting deep within the nebula.

This image was taken across three separate nights from my garden in Cirencester. I used the Ha, Sii and Oiii filters to capture different wavelengths of light and then combined them in post processing. The image is made of approximately 88 5-minute exposures as well as calibration frames.

Elephant's Trunk (IC 1396) wide field

Another image of the Elephant's Trunk Nebula, this time taken with the wider field telescope, the William Optics Z73, at 430mm. The wider field allows you to see more of the detail in the nebulosity.

This photo was taken with the ZWO ASI 1600mm and the William Optics Z73 telescope. It is composed of seven and a half hours of Ha, Sii and Oiii photos combined in the Hubble palette.

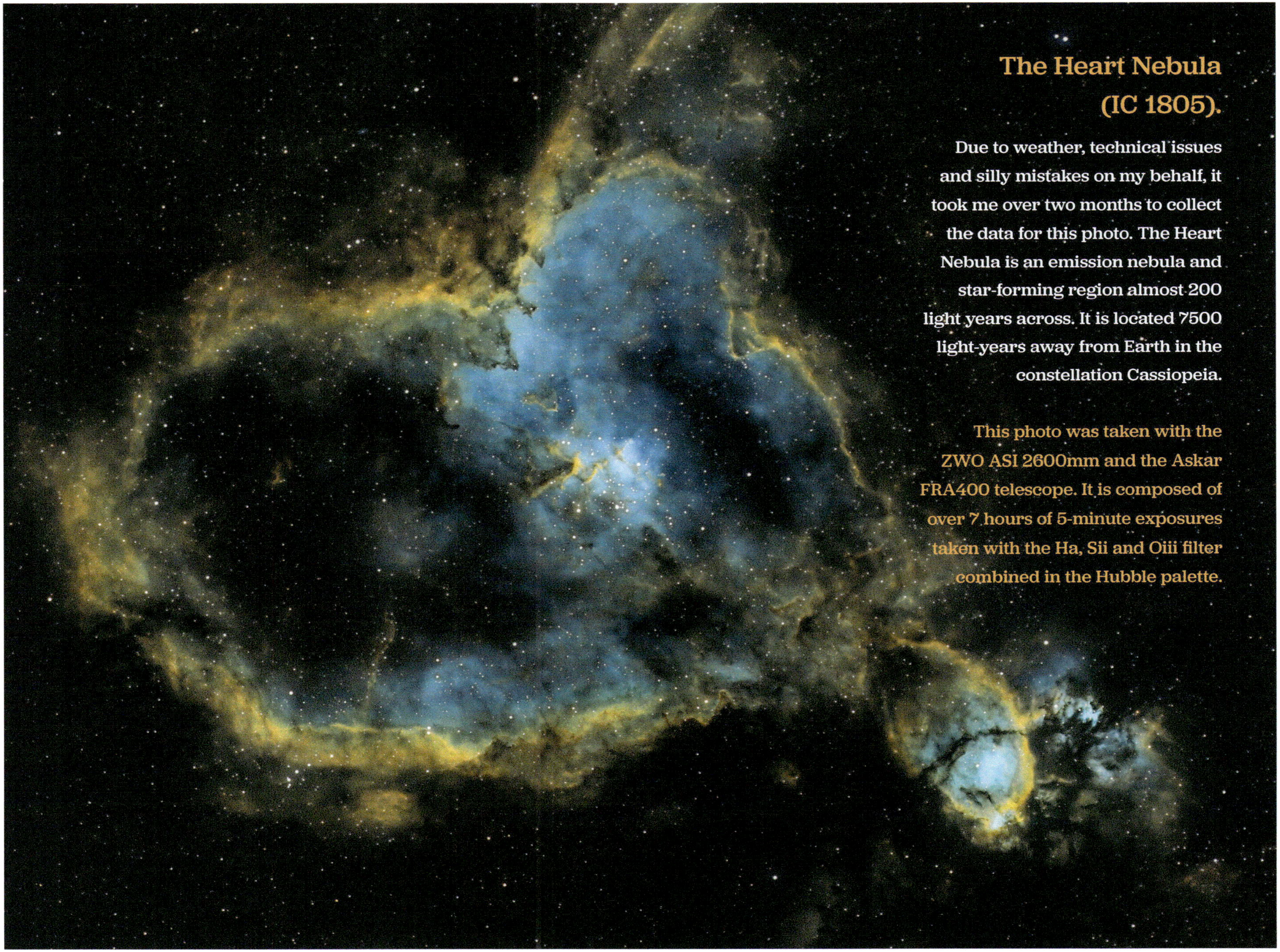

The Heart Nebula (IC 1805).

Due to weather, technical issues and silly mistakes on my behalf, it took me over two months to collect the data for this photo. The Heart Nebula is an emission nebula and star-forming region almost 200 light years across. It is located 7500 light-years away from Earth in the constellation Cassiopeia.

This photo was taken with the ZWO ASI 2600mm and the Askar FRA400 telescope. It is composed of over 7 hours of 5-minute exposures taken with the Ha, Sii and Oiii filter combined in the Hubble palette.

Melotte 15 (IC 1805)

The heart of the Heart Nebula seen on the previous page. The structure in this photo is a giant area of hydrogen gas that is caused to glow by the intense ultraviolet radiation from a young, hot, blue supergiant star that is only 1.5 million years old in the Melotte 15 star cluster. The dust and gas clouds are twisted by the pressure of the intense radiation and the solar wind. This formation is estimated to be 7500 light-years away from Earth, in the constellation Cassiopeia.

This image was taken with the ZWO ASI 1600mm camera and the Skywatcher 190MN telescope. This image contains 11 hours of Ha, Oiii and Sii data collected across five nights.

"ASTRONOMY COMPELS THE SOUL TO LOOK UPWARDS AND LEADS US FROM THIS WORLD TO ANOTHER."

Ralph Waldo Emerson

The Soul Nebula (IC 1848)

The Soul Nebula is an emission nebula in the constellation Cassiopeia. It is located right next door to the Heart Nebula shown on the last page. With a wider field telescope or lens, you can easily capture the Heart Nebula and the Soul Nebula in the same frame. However, with my current equipment I could only capture them in separate images.

This photo was taken across six nights with the ZWO ASI 1600mm and the William Optics Z73 telescope. It is comprised eight hours of Ha, Sii and Oiii photos combined in the Hubble palette.

The Dumbbell Nebula (M27)

The Dumbbell Nebula is a planetary nebula in the constellation Vulpecula at a distance of about 1227 light-years from Earth. A planetary nebula is the type of nebula our Sun will produce when nuclear fusion stops in its core. The nebula is the result of an old star that has shed its outer layers in a glowing display of colour. It's an extremely bright object and relatively close, so it can easily be seen with a small telescope or binoculars from a dark location. This image was captured through the Skywatcher 190MN using the ZWO ASI 1600mm camera.

This image contains three and a half hours of data collected through the Ha and Oiii narrowband filters. No Sii was captured for this image.

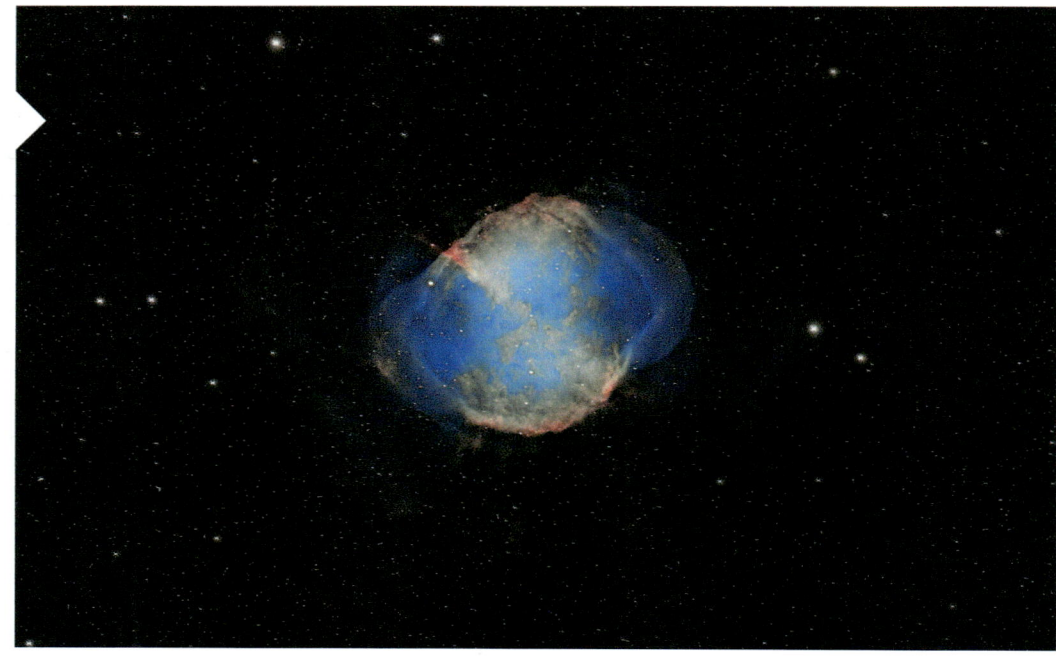

NGC 7538

There is no catchy nickname for this target, but it is an amazing part of the night sky. NGC 7538 is an emission and reflection nebula and a stellar nursery for massive stars. It is an active factory where stars are born, especially huge stars that are over eight times more massive than the Sun. This target consists mainly of hydrogen gas and has a total mass of almost 400,000 Suns. It is located approximately 9000 light-years from Earth. This target is often overshadowed by the more famous Bubble Nebula which lies nearby in the constellation Cepheus.

This image was taken with the ZWO ASI 1600mm and the Skywatcher 190 MN telescope. It contains 167 x 5-minute exposures using the Ha, Sii, Oii narrowband filters.

The Iris Nebula (NGC 7032)

The Iris Nebula (or NGC7023) is a bright reflection nebula in the constellation Cepheus. A reflection nebula is a cloud of interstellar gas and dust that reflects the light from other stars. The nebula gets its blue appearance from one star (SAO 19158), which is responsible for 'lighting it up'. This is a fairly close target as is only 1300 light-years from Earth.

This image was taken with the ZWO ASi 1600mm and the LRGB filters. I used the William optic telescope and captured just over three hours of images split evenly across the LRGB filters.

The Orion Nebula and Running Man Nebula (M42 & NGC 1977)

What can I say about the great Orion Nebula? It's one of my favourite locations of the night sky and probably the most photographed deep sky object around the world. It is one of the brightest and easiest to locate nebulae and therefore one of the easiest to photograph. The Orion Nebula is visible to the naked eye (in dark locations) and lies just south of Orion's Belt, approximately 1,300 light-years from Earth in the consultation Orion. It is an enormous cloud of dust and gas where vast numbers of new stars are being forged. Its bright central region is the home of four massive young stars that shape the nebula. You can just about make out these stars if you look closely at the image. The four hefty stars are called the Trapezium because they are arranged in a trapezoidal pattern. Being only 1300 light-years away, this is the closest large star-forming region to Earth. If you want to start astronomy or astrophotography, then Orion should be one of your first targets. The nebula is seen as the middle 'star' in the 'sword' of Orion, i.e. the three stars that are just below Orion's Belt. This image contains two nebulae, the famous Orion Nebula and the lesser-known Running Man Nebula above. It's easy to see how the Running Man Nebula got its name.

This image was taken with the ZWO ASI 1600mm and the William Optic Z73 telescope. The image contains two hours' worth of 5-minute exposures split evenly through the LRGB filters to capture the detail and colour in the nebula. It also contains 30 x 15 second images for the core. The two images were then blended in processing to create a HDR image.

The Horsehead and Flame Nebula (Barnard 33 & NGC 2024)

The Horsehead and Flame Nebula is located in the constellation Orion, near the most eastern star in Orion's Belt (Alnitak), which is the brightest star in this image, to the left of the Horsehead and above the Flame Nebula. This is a gorgeous part of the night sky and one that is regularly photographed by amateur astrophotographers. With a wider field telescope, you can fit the Horsehead and Flame Nebula and the Orion Nebula in the same image.

This image was taken with the ZWO ASi 1600mm and the RGB filters. I used the William Optic Z73 telescope and captured just over four hours of images split evenly across the RGB filters.

Witch Head Nebula (IC2118)

Another target found in the constellation of Orion, IC 2118 is an extremely faint reflection nebula believed to be an ancient supernova remnant or gas cloud illuminated by nearby supergiant star Rigel. Rigel is a huge blue supergiant that is the brightest star in the constellation Orion and the seventh brightest star in the sky. The nebula glows primarily by light reflected from the star. The Nebula lies in the Orion constellation, about 900 light-years from Earth. The Witch Head Nebula got its name because it looks like the profile of a witch, with the large chin and nose.

While this is a very large target, it is incredibly faint. Even with the 5-minute exposures I was taking for this image, I saw no detail in the nebula. The images just looked completely black, making it extremely hard to frame this nebula. It wasn't until I stacked all of the images and stretched the photo that I was able to see the detail. Due to the faintness, it is almost impossible to view this target visually with a telescope. This image was taken with the ZWO ASI 1600mm and the William optic Z73 telescope.

The total integration time for the final image is approximately 2 hours 30 minutes split evenly between the LRGB filters.

The Bubble Nebula (NGC 7635)

The Bubble Nebula is an emission nebula in the constellation Cassiopeia. The Bubble at the heart of this nebula is 7 light-years across and 7100 light-years from Earth. The bubble itself is caused by stellar winds moving at over 4 million mph and the star forming this nebula is 45 times bigger than our sun. This is one of the most fascinating parts of the night sky and if you want a closer look then you can check out the Hubble Telescope's 26th year anniversary footage, which travels into the heart of the Bubble. After seeing this footage, I knew I wanted to photograph the target myself, and it was one of my first narrowband targets.

This image was taken with the ZWO ASI 2600mm and the Skywatcher 190MN. It contains 105 x 5-minute exposures taken with the Ha, Oiii and Sii narrowband filters.

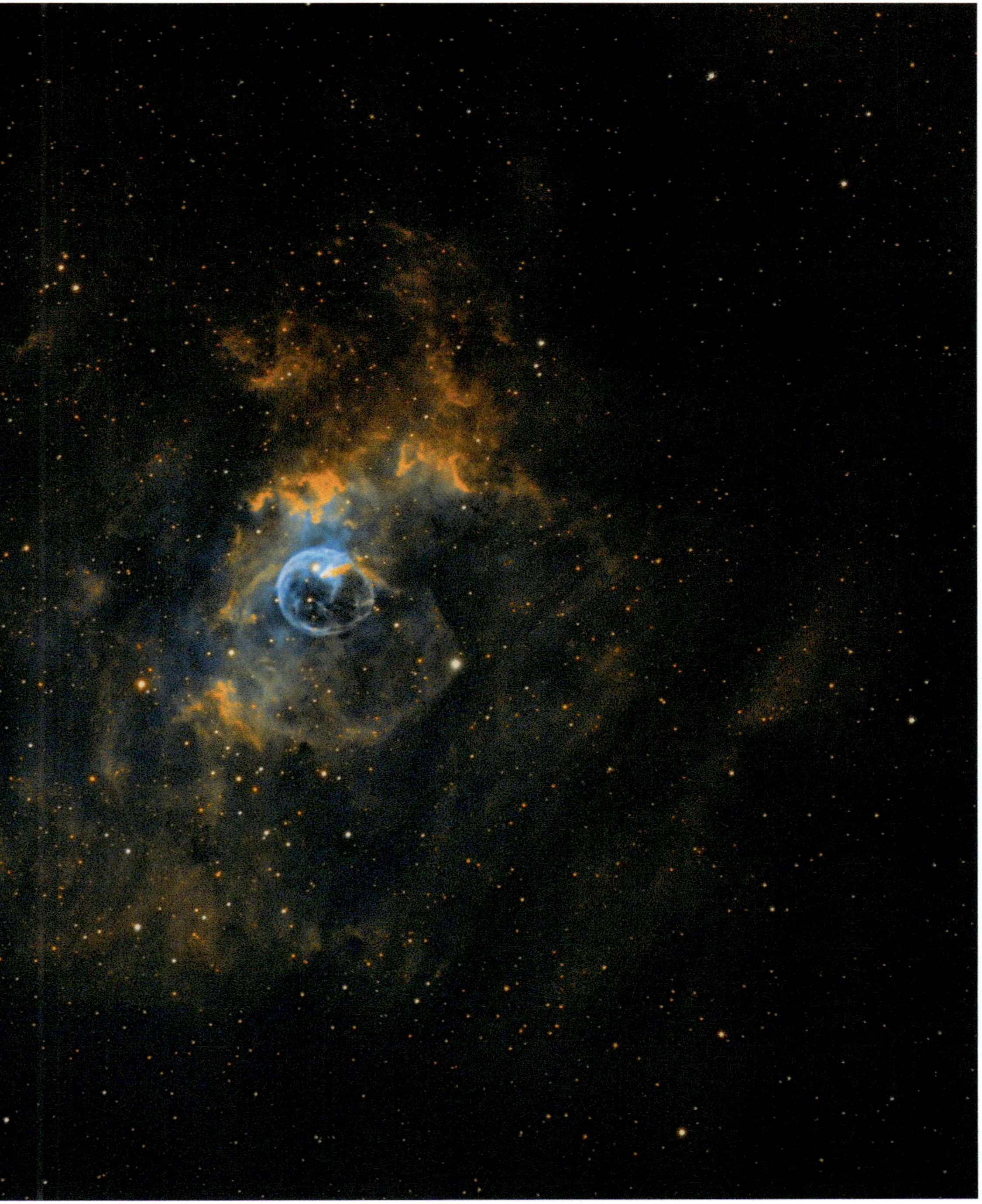

The Lobster Claw and the Bubble Nebula (SH2-157)

Just south of the Bubble Nebula (seen on the previous pages) lies the beautiful Lobster Claw Nebula. This target is a bright emission nebula in the constellation Cassiopeia and, when photographed with a wide field telescope, makes for a wonderful image. Alongside the Bubble Nebula and the Lobster Claw Nebula, there are also numerous other fascinating deep sky objects in this image. The image contains Nova Cas 2021, which is the bright glowing star located to the upper right of the Bubble Nebula. This bright nova became visible over the summer in 2021. A nova is typically caused by a thermonuclear explosion on the surface of a star, however, little is currently known about this nova. Below the Bubble Nebula, you will also see the smaller nebula of NGC 7358, which I have photographed with the larger telescope and which you can see earlier in this book.

This image was taken with the ZWO ASI 2600mm and the Askar FRA400 telescope. The image contains 9 hours 34 minutes of 7-minute exposures taken with the Ha, Oiii and Sii narrowband filters combined in the Hubble palette.

The Pleiades Star Cluster (M45)

Commonly called the Pleiades or Seven Sisters, M45 is an open star cluster in the constellation Taurus, 444 light-years from Earth. The cluster contains over a thousand stars that are loosely bound by gravity, but it is visually dominated by the brightest seven stars or "Seven Sisters". The Pleiades star cluster is one of the nearest star clusters to Earth. It is easily visible to the naked eye and as such one of the most viewed and photographed objects in the night sky. While this target can be viewed and photographed as late as April, it is commonly known as a winter target and gets highest in the sky between September and January.

This image was taken with the ZWO ASi 1600mm and the LRGB filters with the William Optics telescope. I captured just under three hours of images split across the LRGB filters.

"FOR MY PART I KNOW NOTHING WITH ANY CERTAINTY BUT THE SIGHT OF THE STARS MAKES ME DREAM."

Vincent Van Gogh

The California Nebula (NGC 1499)

The California Nebula is an intense emission nebula located in the constellation Perseus, approximately 1000 light-years from Earth. It gets its name as it appears to resemble the outline of the US State of California. This nebula's red glow is caused by the bright star you see in the image, Xi Persei, a blue giant that is over 12,000 times brighter than the Sun. This massive star ionises the hydrogen in the nebula and this is what causes the recognisable shape of this target.

This image was taken with the ZWO ASi 1600mm and the William Optics Z73 telescope. It only contains only just over an hour each of data taken through the Ha and Oiii narrowband filters, so it's not the best quality image you'll ever see. Unfortunately, I only had a quick break in the weather when this target was high in the sky, so I didn't get as much data as I would have liked. I will definitely be revisiting this target next year.

The Flaming Star Nebula (IC 405)

The Flaming Star Nebula is approximately 5 light-years across and 1500 light-years from Earth in the constellation Auriga. The Flaming Star Nebula is a glowing emission and reflection nebula and is lit by the nearby AE Aurigae, a massive star believed to be about 17 times more massive than our Sun and 30,000 times as bright. Interestingly, however, it is believed that AE Aurigae is not originally from this area of the sky or our part of the Milky Way. Astronomers have traced its origins back to the Orion Nebula. It is believed that the star was ejected from the nebula millions of years ago, potentially due to a close gravitational interaction of two multiple star systems in which some stars were flung away from Orion.

    The Flaming Star Nebula is bright enough to be observed through a small telescope and is a pretty easy target to photograph. The hydrogen emission gas makes up the "flame" of IC 405, and the wave-like dust and gas lanes are what give this nebula its name as it looks like smoke being blown from a star which is 'on fire'.

This image was taken with the ZWO ASI 1600mm and the Oiii, Sii & Ha filters. It was photographed through the William Optics Z73 telescope.

The Tadpole Nebula (IC 410)

The Tadpole Nebula is a dusty emission nebula located approximately 12,000 light-years from Earth in the constellation of Auriga. It is right next to the Flaming Star Nebula seen on the previous page, as well as being part of the same star-forming region. With a wide telescope these can both fit into the same image, however with my set-up I need to capture them separately. The nebulosity in this picture is lit by the radiation from the open star cluster (NGC 1893) which lies in the centre of the nebula. The massive, hot stars within this open cluster are all very young, having only been recently formed from IC 410.

    The 'tadpoles' that give this target its name are very similar to the famous Pillar of Creation. They are star-forming regions and very likely to give birth to stars in the future. The stellar winds have caused the shape of the tadpoles, which are pointing their tails outwards, away from the nebula's central regions.

This image was taken over 4 nights with the ZWO ASI 1600mm and the Skywatcher 190MN. It contains 66 x 7-minute exposures taken with the Ha, Oiii and Sii narrowband filters.

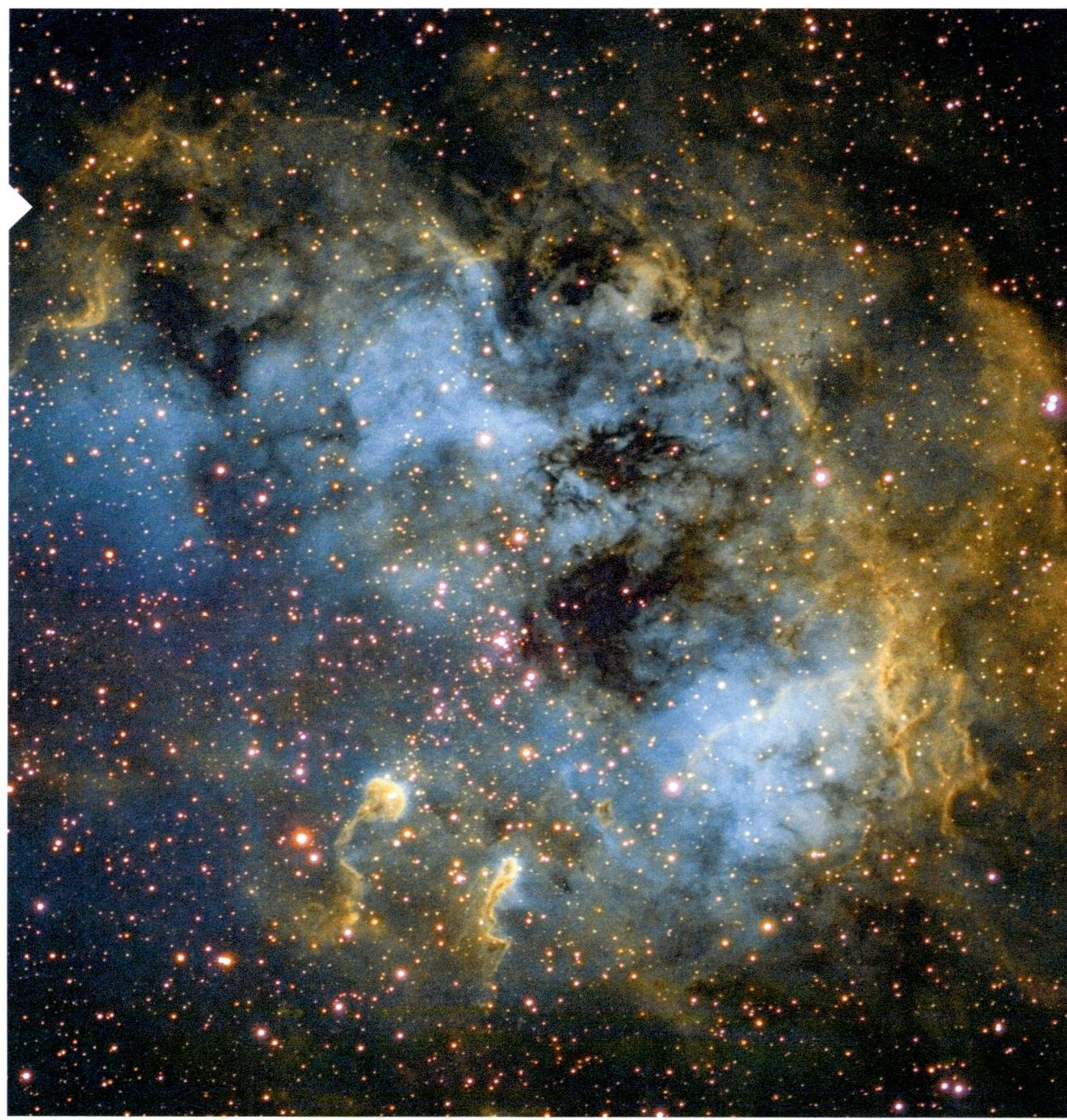

The Pacman Nebula (NGC 281)

This image took exactly one month to complete. I started capturing images on this target on the 4th August and finished on the 4th of September. It is one of my favourites of the images I have captured.

The Pacman Nebula is a cosmic cloud located in the constellation Cassiopeia. It is within our Milky Way Galaxy, approximately 9,200 light-years from Earth. NGC 281 gets its name from the classic video game character. Can you see the resemblance? The nebula is a young open cluster of stars and contains 279 stars (possibly more) which illuminate the gases around it, giving it its distinctive shape.

This image was captured using the ASI ZWO 1600mm camera and the Skywatcher 190MN telescope. In total, this image contains just over 10 hours' worth of images taken through Ha, Oiii and Sii filters. and Sii filters.

On the right, you will see a starless edit of the same image using a special programme called StarNet++, which removes all the stars from astro images.

The Jellyfish Nebula (IC443)

The Jellyfish Nebula, also known by its official name IC 443, is the remnant of a supernova lying 5000 light-years from Earth in the constellation Gemini.

The Jellyfish Nebula is a remnant of a star that exploded as a supernova between 3000 and 30,000 years ago. This beautiful jellyfish was formed when a star that has approximately 8 times the mass of the sun collapsed in on itself, causing a massive explosion. The explosion that caused the jellyfish nebula also created a rapidly spinning neutron star, and astronomers have estimate that this star is shooting through the gas in this image at approx. 800,000 km/h.

This image was taken with the ZWO ASI 1600mm camera and the William Optics Z73 telescope. It took four nights to collect the 6 hours 40 minutes worth of Ha, Sii and Oiii data in this image.

The Rosette Nebula (NGC 2244)

The Rosette Nebula is an open star cluster and an extremely popular winter target. I had to wait patiently for this target to get high enough in the sky for me to photograph, but it was well worth the wait. Although, it is called the Rosette Nebula due to the shape, it is also sometimes referred to as the Skull Nebula. The cluster and nebula lie approximately 5000 light-years from Earth in the constellation Monoceros and measure roughly 130 light-years in diameter.

This image was taken with the ZWO ASI 1600mm camera and the William Optics Z73 telescope. The image contains 44 x 7-minute images taken with the Ha, Oiii and Sii narrowband filters.

"THE COSMOS IS WITHIN US. WE ARE MADE OF STAR-STUFF. WE ARE A WAY FOR THE UNIVERSE TO KNOW ITSELF."

Carl Sagan

The Eastern Veil Nebula (Caldwell 33)

The Easter Veil Nebula is one section of the Veil Nebula, which is a huge area of the sky covering 110 light-years across. The area appears so large that I can't fit it all in the frame, even at a 430mm focal length. It is one of the most well-known supernova remnants (i.e. the remains of a supernova explosion). It is approximately 2100 light-years from Earth in the constellation Cygnus.

This image was taken with the William Optics Z73 and the ZWO ASI 1600mm. The image contains just over 3 hours of Oiii and Ha data.

The Western Veil or Witch's Broom Veil Nebula (NGC 6960)

The Western Veil Nebula, commonly called the Witch's Broom Nebula, is in the same section of the sky as the one above the Eastern Veil Nebula. These areas are so large I just can't photograph with them both in frame, so I have to capture them one at a time.

This image was taken with the William Optics Z73 and the ZWO ASI 1600mm. The image contains just over 3 hours of Oiii and Ha data.

The Wizard Nebula

The Wizard Nebula lies approximately 7200 light-years away from Earth within the constellation Cepheus. It is located pretty close to Cassiopeia's "right-hand" star when looking at the constellation's "W" shape. It is an emission nebula in an open cluster of developing stars, which is surrounded by gases which have formed its wizard-like appearance. Some of the names for deep space objects are difficult to make out, however, the Wizard Nebula is not one of those.

The target rises high in the sky during the summer, making it a summer and autumn target. The issue with photographing summer targets in the UK is the lack of dark night. With just a few hours of darkness each night, capturing narrowband images such as this one is a long process. This image was taken over six separate nights and took nearly two months to complete.

This image was taken with the ZWO ASI 1600mm and the William Optic Z73 telescope. It remains one of my longest-exposed images to date and was taken over six nights with Ha, Oiii and Sii narrowband filters. The total integration time for the final image was 13 hours 45 minutes.

Thor's Helmet (NGC 2359)

The Thor's Helmet nebula is located in the constellation Canis Major and is approximately 15,000 light-years from Earth. The round centre of this nebula and the wing like structures make it resemble the battle helmet wore by Thor, the Norse god of thunder. This emission nebula is approximately 30 light years across, making it far larger than our entire solar system.

This striking nebula gets its glow from WR7, a massive Wolf-Rayet star that many believe will soon turn supernova. Astronomers have calculated that WR7 is 280,000 times brighter than our Sun, 16 times more massive, and 1.41 times as large.

This image was taken over two nights with the Skywatcher 190MN and the ZWO ASI 1600mm. I did not have much time to photograph this target due to its location in the night sky and poor weather. As a result, this image only contains 2 hours and 40 minutes of Oiii and Ha data combined into a bi-colour image. Even though I wasn't able to capture as much data as I had hoped, I am really happy with this final image.

The Crescent Nebula (NGC 6888)

The Crescent Nebula is an emission nebula in the constellation Cygnus, approximately 5000 light-years away from Earth. The nebula is caused by a massive Wolf-Rayet star (a massive, bright, extremely hot star) nearing the end of its life. The star is dying and throwing off a tremendous amount of its mass, causing stellar winds to form this amazing nebula. It is believed that the star will soon (in the next million years or so) go supernova.

This image was taken with the Zwo asi 1600mm and the Skywatcher 190MN telescope. It contains 7 hours of 5-minute exposures taken with the Ha and Oiii narrowband filters. No Sii was collected on this target.

The Cone Nebula and Christmas Tree Cluster (NGC 2264)

Both these objects are located in the Monoceros constellation and are located about 2600 light-years from Earth. Whichever way up you view this target it resembles a Christmas tree. It was therefore the perfect target to photograph the week before Christmas! The Cone Nebula itself is located at the bottom of this image and is a pillar of gas that spans 7 light-years in length.

This image was taken with the Zwo asi 1600mm, the Skywatcher 190MN and the Ha, Oiii and Sii filters. The image contains just over 2 hours 45 minutes of data, so I need to capture some more images on this target to really do it justice. I will have to revisit the Cone Nebula and Christmas Tree Cluster next year.

The Cave Nebula (SH2-155)

SH2-155, also known as the Cave Nebula, is a diffuse emission nebula within a larger complex nebula that includes a reflection nebula, and dark nebula. It is a stellar nursery where stars are born and is approximately 2,400 light years away in the constellation Cepheus.

This image was taken with the Zwo asi 1600mm and the Skywatcher 190MN telescope. It contains just under 14 hours of 7-minute exposures taken with the Ha, Sii and Oiii narrowband filters. This is my longest exposure to date.

Gamma Cygni Nebula or the Butterfly Nebula (IC 1318)

The Gamma Cygni Nebula, often called the Butterfly Nebula due to the shape, is a nebula in the constellation Cygnus, approximately 3000-4000 light-years from Earth. The nebula can be seen next to the bright star of Sadr and the region contains both a dark nebula and an emission nebula, forming the incredibly impressive butterfly shape. The dark nebula makes the body of the butterfly and the emission nebula forms the glowing wings.

This image was taken with the ZWO ASI 2600mm and the Askar FRA400 telescope. The image contains 6 hours, 55 minutes of 5-minute exposures taken with the Ha, Oiii and Sii narrowband filters combined in the Hubble palette.

The North America and Pelican Nebula (NGC 7000 & IC 5070)

This target contains two large nebulae. To the left of the image is the North America Nebula, so called because it is shaped like the continent of North America. Towards the bottom of the North America Nebula you have the famous Cygnus Wall, which I captured with my larger telescope and which you can see on the next page. To the right of the image, you have the Pelican Nebula. While some people say that they can see a pelican in this image, I can't see a pelican no matter how long I look at it. The two nebulae are separated by a large dark nebula, making for a dramatic image. Both nebulae are approximately 1500-2000 light-years from Earth in the constellation Cygnus.

This image was taken with the ZWO ASI 2600mm and the Askar FRA400 telescope. The image contains 10 hours 9 minutes of 7-minute exposures taken with the Ha, Oiii and Sii narrowband filters combined in the Hubble palette.

Cygnus Wall in the North America Nebula (NGC 7000)

The Cygnus wall is the most recognisable and distinctive part of the North America Nebula (seen on the previous page). The wall forms the famous W-Shaped ridge which is approximately 20 light-years in length. The Cygnus Wall makes such a fantastic image that many astrophotographers isolate the wall, and frame it as its own target by using a longer focal length telescope.

This image was taken with the ZWO ASI 2600mm and the Skywatcher 190MN telescope. The image contains 7 hours 25 minutes of 7-minute images across the Ha, Oiii and Sii narrowband filters combined in the Hubble palette.

Eagle Nebula and the Pillars of Creation (M16)

What can I say about this image? It contains the infamous Pillars of Creation, which were made famous by probably the most iconic Hubble Telescope images of all time. If you ask someone to think of a NASA image, they will probably picture the Pillars of Creation. I was desperate to photograph this target, but it is in a very inconvenient position in the night sky for me. It stays very low to the southern horizon and is blocked by both my house and the neighbour's tree. Therefore, I only have just over an hour each night to capture data on this target for just a few months of the year. For this reason I couldn't capture as many images as I had hoped, but I was still delighted to be able to photograph the famous pillars. The Eagle Nebula itself is approximately 7000 light-years away in the constellation Serpens, with the Pillars of Creation making up a small (4-5 light-years) feature within the larger nebula.

This image was taken with the ZWO ASI 2600mm, the Skywatcher 190MN telescope. The image contains only 4 hours 45 minutes 5-minute exposures taken with the Ha, Oiii and Sii narrowband filters combined in processing to the Hubble palette.

The Tulip Nebula (Sh2-101)

The Tulip Nebula is an emission nebula found in the Cygnus constellation, approximately 8000 light-years away from Earth. While it takes a lot of imagination to see how some deep sky objects get their name, this is not the case with the Tulip Nebula. This 70 light-years-across flower in space is such a gorgeous target and is one of my favourite images to date. The Tulip Nebula is located relatively close to the Crescent Nebula and when photographed with a very wide field telescope you can just about fit both in frame together, something which I hope to do in the future.

This image was taken with the Skywatcher 190MN telescope and the ZWO ASI 2600mm. It contains 6 hours 50 minutes of 5-minute exposures across the three Ha, Sii and Oiii narrowband filters combined together in the Hubble palette.

The Lion Nebula (Sh2-132)

The Lion Nebula is another deep-sky object that takes little imagination to see how it received its name. Located in the constellation Cepheus approximately 10000 light-years from Earth, this amazing lion-shaped nebula is often overlooked by astrophotographers. I'm not sure why more people do not photograph this target. Maybe it's because there are lots of other fantastic targets in the night sky this time of year, or maybe because it's because it lies right next to the very famous Elephant's Trunk Nebula. Either way, I'm glad that I attempted to capture this faint target. Even though I photographed this target across two nights with a 99% moon, I'm happy with how the image turned out.

This image was taken with the Askar FRA400 and the ZWO ASI 2600mm. It contains 8 hours 45 minutes 5-minute exposures across the three narrowband filters (Ha, Oiii, Sii) combined in the Hubble Palette.

The Crescent Nebula Collaboration (APOD winning photograph)

Nearly a year after photographing the Crescent Nebula, I was contacted by two fellow astrophotographers, Joe Navara (www.joesastrophoto.com) and Glenn Clouder (www.astrobloke.com). Like me, they both have their own astrophotography YouTube channels and they asked me if I wanted to participate in a collaboration image. After a long discussion, we decided that the best target we could all shoot was the Crescent Nebula. We all set about photographing the same target and, after a few weeks, we got together to edit the image. Between us, we managed to collect over 30 hours of data on the Crescent Nebula, all using our different telescopes, mono astro cameras, and Ha and Oiii narrowband filters to combine into a HOO image. After a two and a half hour Zoom meeting, we had our final image and we were all extremely happy with the final result. We hoped to end up with a nice image due to the large amount of data, but we were all amazed with the detail in the nebula and especially the amount of detail we were able to reveal in the oxygen regions on the nebula.

Because of this, Joe decided to submit the image to NASA for the Astronomy Picture of the Day (APOD). To our amazement, NASA selected our image for the APOD, on the 17th of June 2021. It is absolutely incredible to think that our image was published by NASA and will forever be in their image archives alongside images captured by professional state-run organisations and famous telescopes such as Hubble. The image was seen around the world, aided by a BBC article that was published on the image.

This image is the most recent target I photographed before this book was published and I am truly blown away that it was selected for a NASA APOD. Having only started this hobby at the beginning of the COVID-19 lockdown, I never dreamt that I would have an image published by NASA. I'm so thankful to Joe and Glenn for inviting me to collaborate and for this amazing hobby, which has been so rewarding and helped keep me sane during what has been such a terrible year.

Comet Neowise – from Cotswold Water Park (left) and from home (below)

Between the first and second lockdown, the easing of restrictions meant I was able to travel locally to capture some landscape astro photos. In July, we were treated with a very special visitor in Comet Neowise. This became the brightest comet since comet Hale-Bopp in 1997 and as such became the most photographed comet of all time. I could not resist the chance to capture this once in a lifetime comet.

The image to the left was taken with the Sony a7iii and a wide angle Zeiss 25mm lens in the beautiful Cotswold Water Park, just 5 miles from my home. It contains 27 photos stacked, f2.0, 10 sec, ISO 500. The image above was taken using the longer 100-400mm lens at 297mm from my garden in Cirencester. Settings were f5.6, 20 sec, ISO 400.

Comet Neowise - The Rollright Stones

I took a short 45-minute drive out to the dark sky site of the Rollright Stones, an arrangement of megalithic monuments located between the Oxfordshire and Warwickshire border. The monuments span nearly 2000 years of Neolithic and Bronze age development and each of the multiple monuments dates from a different period. There are three main monuments in total and the site offers wonderful views of the night sky. The site, which is on the edge of the Cotswold hills, was designated a dark sky discovery sight in 2014. While there are much darker skies in the UK, this is definitely the darkest location close to my home and at only 45 minutes away, I will definitely head back when the lockdown restrictions are lifted. Being a dark sky discovery site, the Rollright Stones regularly feature on TV programmes such as *Stargazing Live* and *The Sky at Night*.

The image to the right shows Comet Neowise framed between two trees. In the top of the image, you can see the constellation of Ursa Major, also known as the Plough or the Big Dipper. In the foreground, you can see one half of the Kings Men stone circle. This ceremonial stone circle was erected in approximately 2500 BC. There are currently 70 stones which have survived the test of time.

This image was taken with the Sony a7iii camera Zeiss 25mm F2. It contains 14 photos stacked f2.2, 13 sec, ISO 1600.

On the next two pages, you will see a collection of Milky Way images taken at the Rollright Stones in the same night I captured this image of Comet Neowise.

Sony a7iii, Zeiss 25mm f2.0 43 photos stacked f2.0, 10 sec, ISO 1600

Sony a7iii, Zeiss 25mm f2.0 25 photos stacked f2.0, 10 sec, ISO 1600

Sony a7iii, Rokinon 14mm f2.8 10 photos stacked f2.8, 20 sec, ISO 2000

Sony a7iii, Rokinon 14mm f2.8 13 photos stacked f2.8, 20 sec, ISO 2000

I took a trip to out to Rodborough Common, approximately 25 minutes from our home, to get some images of the Milky Way. I managed to get two images from the same location, one with car headlights and one without. Which do you prefer? Both images were taken with the Sony a7iii, Rokinon 14mm f2.8. Both images contain 22 x 20 second images shot at f2.8, ISO 1600. The image on the right has one additional image taken to include the car headlights and blended in Photoshop to create the foreground.

The Milky Way Arch - Cotswold Water Park

I feel incredibly lucky to live extremely close to the beautiful Cotswold Water Park, which covers over 40 square miles and has over 180 lakes. We live right next to the heart of the park where the majority of lakes are based. I visit the park all year round for dog walking, photography and just to admire the beautiful area. One of my other passions is wildlife photography and the park is home to a huge range of species from rare birds to otters and deer.

Of course, the lakes also offer a great chance to take wide-angle astro photos. I just love the reflections of the stars in the water that you are able to capture if the lake is still. The site itself is not the darkest location ever and you get a lot of light pollution from nearby Swindon, however, you are still able to pick out the Milky Way in astro images.

The image on this page is my first attempt at capturing the Milky Way arch. It was captured with the Sony a7iii, Zeiss 25mm. It is an 11 shot vertical Panorama image. 25mm, f2.0, 15 sec, ISO 1600.

"WHEN I CONSIDER HOW, AFTER SUNSET, THE STARS COME OUT GRADUALLY IN TROOPS FROM BEHIND THE HILLS AND WOODS, I CONFESS THAT I COULD NOT HAVE CONTRIVED A MORE CURIOUS AND INSPIRING SIGHT."

Henry Thoreau, Poet

This is probably my favourite astro landscape image taken in 2020. I love the reflection of the tree in the middle of the image and the reflection of the bright stars in the water. Taken with the Sony a7iii, Rokinon 14mm f2.8. The image contains 22 photos stacked in post processing. f2.8, 20 sec, ISO 1600

The Milky Way – Uley Bury Hillfort

Uley Bury Hillfort is one the finest example of a hilltop fort in Gloucestershire. It is a large iron age fort dating to around 300 B.C, which offers stunning views of the Gloucestershire countryside and beyond. The fort is just outside of the small village of Uley, due to the lack of light pollution in the area, and the stunning views to the South, it offers astrophotography's a great opportunity to capture some Milky Way images.

The image on this page was captured with the Sony a7iii, Zeiss 25mm. The image contains 13 photos stacked in post processing. f2.0, 10 sec, ISO 2500.

"THE REAL FRIENDS OF THE SPACE VOYAGER ARE THE STARS. THEIR FRIENDLY, FAMILIAR PATTERNS ARE CONSTANT COMPANIONS, UNCHANGING, OUT THERE."

James Lovell, Apollo Astronaut

This was captured with the Sony a7iii, Zeiss 25mm. The image contains 16 photos stacked in post processing. f2.0, 10 sec, ISO 2500.

The above was captured with the Sony a7iii, Zeiss 25mm. The image contains 21 photos stacked in post processing. f2.0, 10 sec, ISO 1600.

Milky Way in the Cotswolds

The rolling hills of the Cotswolds offer some fantastic locations for Milky Way photography. I constantly find myself scouting for new locations and checking how dark different areas are. During the early summer of 2021, at about 1am one morning, I took a short trip up the road to a little village called Turkdean, a small village approximately 5 miles south of Bourton-on-the-Water. I was able to find three compositions that I was happy with on this trip, which can be seen on the next few pages. While there was more light pollution than I hoped from Stow-on-the-Wold and Bourton-on-the-water, I was still happy with these images.

The image on this page was captured with the Sony a7iii and the Sony 14mm f1.8 lens. The image contains 31 photos stacked in post processing at f1.8, 15 sec, ISO 1600. I also took a longer exposure of 175 seconds at f2.8, ISO 640 for the foreground and combined the image for the sky and foreground in Photoshop.

The image on this page was captured with the Sony a7iii and the Zeiss 25mm. The image contains 34 photos stacked in post processing at f2.0, 10 sec, ISO 2500. I also took a longer exposure of 154 seconds at f2, ISO 640 for the foreground and combined the image for the sky and foreground in Photoshop.

The image on this page was captured with the Sony a7iii and the Sony 14mm f1.8 lens.
The image contains 24 photos stacked in post processing at f1.8, 15 sec, ISO 1600.

Malvern Hills - British Camp

On another clear night in the summer of 2021, I took a trip to the Malvern Hills. It's a place that I have not visited before, even though I only live 45 minutes away, so when a fellow photographer invited me on a Milky Way photo shoot I jumped at the chance. We decided to go to the British Camp, which is an Iron Age hill fort in Herefordshire. My friend promised me a gentle 10-minute walk to the top and, while it is pathed the whole way up, I was certainly out of breath after the 1000+ feet climb. From the top of the hill there are amazing views of the countryside (not that we could appreciate them in the dark) and you are high enough to see the Milky Way core above the landscape. I captured two images from the top which I was happy with, and you can see these here.

The image on this page was captured with the Sony a7iii and the Sony 14mm f1.8 lens. The image contains 28 photos stacked in post processing at f1.8, 15 sec, ISO 2000. I also took a longer exposure of 142 seconds at f1.8, ISO 640 for the foreground and combined the image for the sky and foreground in Photoshop.

The image on this page was captured with the Sony a7iii and the Sony 14mm f1.8 lens. The image contains 35 photos stacked in post processing at f1.8, 15 sec, ISO 2000. I also took a longer exposure of 131 seconds at f1.8, ISO 640 for the foreground and combined the image for the sky and foreground in Photoshop.

ISS Moon Transit

The International Space Station flies 400 km high at 15,500 mph. It only takes 92 minutes to make a complete orbit of Earth, so it goes around the planet every 92 minutes! Astronauts working and living on the ISS experience 16 sunrises and sunsets each day.

This image was taken with the Sony a7iii and the Sony 200-600 lens. It consists of 20 images that were taken within a 2-second window stacked together in Photoshop.

BV - #0177 - 160523 - C70 - 216/280/5 - PB - 9781739864552 - Matt Lamination